A New Look at the Mechanisms and Theory of Aging

A NEW LOOK AT THE MECHANISMS AND THEORY OF AGING

LEV SALNIKOV

To order additional copies of this book, contact:
Xlibris Corporation
1-888-795-4274
www.Xlibris.com
Orders@Xlibris.com
94916

CONTENTS

Possessing me thou shalt possess all things. But thy life is mine, for god has so willed it. Wish, and thy wishes shall be fulfilled; but measure thy desires, according to the life that is in thee. This is thy life, with each wish i must shrink even as thy own days. Wilt thou have me? Take me. God will hearken unto thee. So be it!

—*The Magic Skin*, **Honoré de Balzac**
(*La Peau de Chagrin*, Honoré de Balzac)

From the Author

One of the great mysteries yet unsolved by the modern biological science is the cause of aging of every multicellular organism known to science, including *Homo sapiens*. This issue is ever more relevant as the causes underlying the majority of human diseases have aging processes at their core. Many years of research and an immense number of scientific works and theories dedicated to this subject have failed to yield a conclusive answer.

Indeed, if we speak of the animal kingdom, it seems that every species had been dealt a certain amount of time to live. Regardless of external causes or conditions, "the magic skin" of the life force allotted by nature inevitably gets smaller, faster for some and slower for the others, but inevitably leading to aging and death. The present work is an attempt to unravel just what is this effect of "the magic skin."

The author believes that a *new look* at, or an alternative view of, the mechanisms of aging is no less important than any experimental work in this area. Very frequently, it is an unbiased examination and a fresh perspective on the existing problem that helps to finally reach its resolution. The author tried here to present the material without undue complications while avoiding oversimplifications. This may well earn him reproach on the part of the specialists; however, it is important to keep in mind that any thought, or the absence thereof, can be hidden behind impenetrable terminology and an academic presentation style. The author hopes that by using a more accessible presentation style, he will be able to reach a broader audience interested in the general problems of biology.

The author is aware that he does not possess the expertise in all areas of biology associated with the problems of aging. Nevertheless, it is quite likely that such people simply do not exist. Of course, the best way to avoid criticisms and errors is simply to write nothing at all. Yet by doing nothing, it is impossible to achieve a result. It is well-known that only by expressing and considering a variety of opinions is it possible for us to move in the direction of the truth.

Thanks to the modern information technologies, it no longer takes a great deal of effort to gain access to the experimental results in most areas of science. Fortunately for the author, gerontological research is no exception, allowing him to avoid engaging in the experimental work but to simply participate in the discussion of the comprehension of its results. In summarizing the currently available data about the phenomena associated with the aging processes in the multicellular organisms, we can safely say that their basis is in the steady decline of the biosynthesis and the reparative abilities, which commences subsequently to the organisms' sexual maturation. This naturally leads the organism to the majority of known age-related changes.

However, even today, despite a huge number of research studies aimed at the questions of age-related changes, the specific causes or mechanisms leading all members of the animal kingdom to their eventual demise have not been identified. In many respects, the present state of affairs in this area is beginning to resemble the situation of "the search for the black cat in a dark room." It seems that it is simply impossible to identify a universal cause of aging; however, as Oscar Wilde had put it, "Nothing is worth doing except what the world says is impossible."

Any new hypothesis or theory devoted to the mechanisms of aging must answer the key questions—*why* and *how*. In order to attain the desired result, it is necessary to once again take a good, unbiased look at the well-established facts. In effect, the author presents here an attempt to overcome the "hypnotic effect of the certainty" and to reconsider the established facts anew while combining them with respect to a certain new perspective.

Niels Bohr had once said that the inability of the biologists to understand the essence of life stems from their two classical approaches—observation and deconstruction—being mutually exclusive and that biology would encounter paradoxes similar to those that already exist in quantum physics. Indeed, the choice of the basic axioms, or the initial postulates, being used in considering the fundamental problems in biology is very relevant today. It is not sufficient to simply record the experimental results and established facts. The most important thing is their correct understanding and interpretation, which leads to the creation of a self-consistent description of the phenomenon in question.

Accordingly, the goal of the present work is to offer the author's own version of the causes and mechanisms of aging in multicellular organisms. Here, we will consider the process of aging of a multicellular representative of the animal kingdom, which in itself presents the most interest to us. Such organism will be taken as the general object of the description and analysis in this work.

Along the way, the reader will be presented with the analysis of the current theories of aging and the analysis of the role of ontogenesis in terms of the evolutionary processes. In addition, the relationship of the organism with its constituent cells and the implementation of the ontogenetic programs and their role in the aging processes will be discussed. Naturally, the theory, or the hypothesis, being offered here, as with any other, must do more than explain the already known, more universally accepted facts—in this case, the relationship between the rate of sexual maturation and the average life span and the increase of the life span with the reduction of the metabolic rate. The explanation of the mechanisms associated with the DNA damage inflicted by the free radicals, the physiological processes of neurohumoral deregulation of metabolism as a function of aging, the meaning and the role of the telomere clock, etc., must also be presented here. Will the approach being offered by the author lead to a new result? We will see the answer to this question at the end of this work.

The Current State of Understanding of the Causes and Mechanisms of Aging

At present, there is a vast amount of literature dedicated to the subject of aging. To cover all of it is simply impossible, and it is neither the goal of this text. A curious reader can easily find this information on the Internet. In order to introduce the currently available ideas pertaining to the process of aging, we will examine the key theories from the standpoint of their foundation or, in other terms, from the standpoint of what has been *postulated* by their authors and followers.

We will begin from the, chronologically, first approach to thinking about aging, which we can designate as the *wear and tear* view. The main postulate, which makes up the foundation of this group of theories, is the *inherent inability* of the biological systems to oppose the damage due to stochastic chemical processes. The most well-known theory of aging based on this postulate is the *error catastrophe* theory, which states that the cellular DNA accumulates the errors resulting in translational error leading to loss of proper protein function. As a result, the cell accumulates mutations, and the number of dysfunctional proteins increases (1-23).

By postulating the inherent imperfection in the construction of the living organisms, the authors of this theory are forced to ignore, for example, the fact of existence in nature of bacterial species that have lived for eons while their genomes remain practically unchanged. Inherent imperfection of the organisms is also postulated by the theories that explain aging as the product of the inevitable errors that occur in the process of the normal operation of any complex system. The authors find the basis for such conclusions in the mathematical models related to the theory of reliability (24, 25). Still, what is overlooked here is the more basic difference of the living systems—the ability to repair the damage. The theories of aging, which point to the degenerative processes in the various tissues within the organism such as neuroglia (26, 27), etc., can also be grouped into the same category. However, here, it is also completely unclear what constitutes the trigger mechanism for such changes in the tissues. Many other hypotheses can be listed here in which the disruption of functionality of various parts of the cell—mitochondria and cell or nuclear membranes—are identified as the causes of aging.

As we will see later, just as in the case of the hypotheses that propose the disruption of the functions of the various tissues and systems in the organism as the triggering mechanism, the actual cause that sets in motion the process of aging remains outside the framework of any such hypothesis. All components of the cell, as are all cells in the organism, are linked and interdependent, giving rise to a unified system. It can be argued that all component of such systems are equally important and are subject to coordinated changes in the course of the aging processes. This situation leads to a condition under which the disruption of the function of virtually any part of such a living system may act as the trigger factor of the aging processes.

The theory of aging that extends the key role of the aging processes to the cellular mechanism of *apoptosis* suffers from the same deficiency (31, 32). The fact that the mechanism of apoptosis is present in every cell of the organism in no way provides an explanation for the causes of its sudden activation. Reference to the accumulation of the cellular DNA damage associated with the side effect of oxidative-reductive processes of the cell simply shifts the cause of the aging process onto the same imperfection of the living organism and, consequently, wear and tear.

An alternative to the aforementioned theories is another large group of theories of aging, which is usually referred to as the *evolutionary theories*. We shall assign them to the group of theories that postulate the presence in the organism of a genetically predetermined *mechanism of death* or *program of aging*. These theories propose that in the course of evolution, the mechanisms of aging have evolved along with the appearance of the multicellular organisms as a necessity for their ability to evolve further.

Within this framework, it is worth noting a few works aimed at attempts to identify the "death gene." In this case, the genome is attributed with having a predetermined program of assured destruction of the organism, a sort of "genetic bomb," or a programmed disruption of the normal gene functions in the cell. Such views of the causes of aging are presented in the classical works of Weissmann, Medawar, Williams, and others (33-46).

According to a different theory based on the phenomenon of evolution and known as the theory of antagonistic pleiotropy (46-57), aging occurs because natural selection contributes to the consolidation of the alleles, which have a favorable effect early in life, even if they are unfavorable later on. Thus, the selection contributing to such allele is realized only in the formative period and is indifferent toward the function of the cell's genes later in life of the individual. This situation, according to the author and supporters of this approach, leads to a cascade of error and damage accumulation in the genetic machinery and constitutes the cause of aging of the multicellular organisms. However, until now, it has not been possible to identify a gene or a group of genes that, when naturally or artificially mutated, would change the life span of the species by several fold.

Attempts to identify a specific genetic mechanism responsible for aging include the work devoted to the function of the chromosomal machinery of the cells. That is how the limit of the number of cellular divisions, named the Hayflick limit, after its discoverer, was discerned (58-62). In the later work, the relationship of this phenomenon with the chromosomal telomeres had been established (63-69). In both cases, the authors propose the mechanism that limits the number of cell divisions to be the cause of aging. A major shortcoming of this explanation of aging is, firstly, the presence in the body of stem cells, which have no such limit on the number of divisions they can undergo at all ages. Secondly, this theory applies only to the dividing cells, leaving out the postmitotic cells that form the basis for the functional activity of the organism.

The group of the *evolutionary theories* should also include the theories that accept the simultaneous existence of multiple mechanisms of aging—first, the more ancient universal mechanism, and second, particular to the specific species. As an example, consider the salmonids for which, as for some other fish, death is associated with hormonal processes brought about by sexual maturity (spawning). However, it is known that the removal of the gonads from the members of these species leads to the prolongation of their life span to values close to their neighboring species. From the mechanisms of self-destruction described in the literature, it is necessary to underscore the evolutionary theory of aging (70, 71).

According to this theory, the mechanisms of aging come down to the continuous elevation of the sensitivity threshold of the hypothalamus to the hormonal levels in the blood. Such changes in the regulation lead to an increase in the concentration of hormones circulating through the body. Consequently, the physiological status of the organism is disrupted in a number of ways, in turn, leading to the development of chronic illnesses, aging, and finally death. The key postulates of this theory are the following:

1. Developmental program of the organism requires programmed changes (disturbance) of the level of homeostasis.
2. In mammals, changes in the sensitivity threshold of the hypothalamus to the regulatory signals play a critical role in the mechanism generating the required disturbance of homeostasis during postembryonic development.
3. Conservation of this mechanism, following the completion of the development, is directly responsible for the transformation of the program of development into the mechanism of aging and the major diseases associated with it.

Following the author's logic, the program for *ontogenesis* of the organism also contains within it a *program of aging*. Thus, in the organism, the main role in the initiation of the aging process is delegated to the neuroendocrine regulation and the cells that carry it out. At the same time, it remains somewhat unclear what role in the mechanism of aging described here by the author is reserved for the ontogenetic program and what role is played by the accumulation of errors in the neurons of the hypothalamus due to the damage caused by free radicals. In other words, it is unclear what initiates this program—age-related damage of the neurons in the hypothalamus or a separate, genetically predetermined mechanism related to ontogenesis (72).

One of the modern theories of aging, which is an attempt to unite the majority of the currently known facts and viewpoints, is the *theory of the disposable soma*, put forward by Thomas Kirkwood (73-75). Let us examine it in some detail. According to Kirkwood, aging is a result of the *progressive limitations* on the energy expenditure of the organism on the repair of the *soma* (the somatic cells are all cells of the organism with the exception of the gametes), which arise as a result of the competition for the carbohydrate and protein resource with the reproductive processes.

In Kirkwood's opinion, the key processes that determine the life span and the rate of aging are controlled by the genetic determinants, which regulate supporting functions such as DNA repair genes, antioxidant enzymes, and stress proteins. It is argued that the process of aging is, by its nature, stochastic, but that the life span as a whole is *preprogrammed* by the aforementioned genes. Kirkwood argues that the maximum life span is not defined as a clear time frame but can be changed, for example, by varying the ratio between damage and the gain of function of soma maintenance.

Many of the earlier theories of aging consider the various hypotheses as competing. In contrast, the theory of the disposable soma suggests that the various types of damage accumulate in the cells *in parallel*. The theory of the disposable soma claims that in order to increase longevity, it is necessary to invest more energy into the *maintenance of the soma*, which correspondingly decreases resources available for *reproduction*. According to Kirkwood, the resource reserves of the organism are limited, and consequently, they are redistributed between the expenditure on the reproduction and the maintenance of the somatic tissues. This situation stems from the fact that there are *programs* for the support of the repair functions, and at some point in life, these functions either deteriorate or cease altogether.

A question arises, however, on the level of genome: what regulates the activity of the repair genes? We end up in the same logical trap with the same question: who governs the governor? Unfortunately, the author of this theory evades the direct question of the primary causes of aging by equating them to the presence in the genome of the specific programs of aging.

Concluding this section, we see that the scientific understanding of the nature and causes of the phenomenon of aging of multicellular organism contains two main postulates presented above: inherent imperfection and the presence of the program of aging. The research in these areas continues at the present time.

All attempts to explain the phenomenon of aging using the postulate of wear and tear as a consequence of the inherent imperfection run into the necessity to find a single key link that can be controlled to regulate the processes of aging or stop it altogether. To date, the contender for the role of such a link is the inability of the cell to repair the damage to its own genome completely. That is the conclusion produced by the experimental work. For example, when a result is obtained demonstrating an increase in the damage of DNA in the old cells relative to the young cells, an automatic assumption is made that this is the cause of aging. A question arises however: could this result be a consequence of some other process? After all, there is a well-known rule of logic that *following* does not always mean "as a result of." So the products of other processes may be mistaken for the cause of the phenomenon.

Following such logic, it seems possible to consider infinitely more elementary levels of organization, from the molecular down to quantum mechanical, and still not find the answer to the main question of the cause of aging. However, it is worth noting that the paradigm of the imperfection remains pretty attractive as it provides an illusory ability to fix the errors of nature and to find the link in the chain of events, which can be corrected to obtain a perfect and, therefore, immortal organism.

The second postulate, namely the presence of the program of aging in the multicellular organisms based on the evolutionary necessity of aging as a mechanism of the generational changeover, is undeniably justified and attractive. By denying the existence of any sort of weak links in the cells and in the organism and taking into equal consideration the importance of all its parts, this approach naturally leads to the conclusion of a special program directed at aging. Though let us reiterate that up until now, this program that would allow to instantly alter the life span of the experimental subject has remained elusive.

Note that due to the disregard of evolution toward the individual after it reaches sexual maturity, we cannot conclude that this disregard alone will, inevitably and for each species, launch the process of aging. So as an example, let us consider the elevation theory, which directly indicates the presence of a special mechanism that initiates the aging processes. In addition, as in the case of the wear and tear postulate, the postulate of the presence of a specific program of initiation of aging suggests the possibility of the ability to control it and to turn it off, which in fact is the ultimate goal.

Everything discussed here leads to the conclusion that a hypothesis or a theory that not only directly points to the proposed mechanism of aging but could also be tested experimentally does not yet exist. This, in effect, is the main problem from the viewpoint of the present work. One of the paths to the solution of this problem is the analysis of the currently available data based on the fundamental properties of organization of the multicellular organisms, their construction, and their phylogeny. Based on such an analysis, it is possible to create a new approach to the description of the mechanisms of aging, an alternative to the initial orientation toward the direct observation or the result of deconstruction. The path to the solution of this problem is to solve the puzzle from the existent, already-known data but from a different viewpoint.

Evolution and the Phylogeny of the Multicellular Organisms

From the first part of this text, devoted to the overview of the currently existing views on the causes of aging-related changes, we can conclude that the most commonly accepted idea supported by experimental evidence is that aging is a result of the progressively increased limitations placed on the energy expended on the maintenance and *repair* of the somatic cells. Such limitations are temporally correlated with the sexual maturation and the cessation of the growth phase of the organism.

In this section, we will consider the place that the multicellular organisms occupy in the biosphere and the possible scenarios for their appearance in the course of evolution, as well as their role in the processes of interactions within ecosystems. We will determine whether there is a tendency toward increased complexity of the organization of the living systems in the biosphere, what is the connection between ontogenesis and evolution, and what remains constant as evolution progresses. In conclusion, we will try to identify the location of the aging processes in the multicellular organisms, as well as their *necessity* or *inevitability* from the standpoint of evolution. In other words, we will consider the question of whether a mechanism of death, which regulates adaptational and reparative abilities of the cells of the multicellular organisms, had developed as a result of evolution.

It is a well-known axiom that ontogenesis recapitulates phylogenesis. The relationship of the processes of evolution and ontogenesis (in the case when the program of ontogenesis is responsible for the aging) in the multicellular organisms within a biosphere is evident even from the fact that without the changeover of the generations, the process of evolution itself is not possible. Let us consider in more detail the process of appearance of the multicellular organisms or the process of their *phylogenesis* in the course of evolution.

As early as 1859, Charles Darwin formulated the conditions necessary for the evolutionary process. It is the triad encountered in the theory of evolution—*heredity*, *variability*, and *natural selection*. The meaning of the word *heredity*, as it is used here, is self-evident. However, *variability* and *natural selection* deserve a closer examination.

Heredity and variability are two sides of a *single* process related to the continuous replication of the organisms of any degree of complexity in nature. Traditionally, the basic functional unit of evolution is a population, a relatively isolated number of genetically similar individuals. The population can be considered as a "super-individual" made up of its own genetic copies. Such way of life provides an immense adaptational advantage relative to the hypothetical individual represented by a single unique genetic copy. Consequently, the production of these copies, a process based on a continuous turnover of the generations, is the *key prerequisite* for the process of evolution itself. The production of copies in the unicellular organisms is simply the process of cell division. Naturally, the process of division in the multicellular organisms based on replication and turnover of generations, while it appears different, serves the same purpose—adaptation, based on qualitative change. In other words, the adaptational responses of the organism change in the course of evolution. It should be emphasized that such an adaptation is only relevant for the structural and functional features of the somatic cells, which, in effect, serve as a shell for the germ cells, ensuring their continuous reproduction.

It remains an open question what the mechanisms are for the acquisition of new genetic information in the biosphere. A detailed description of the methods for attainment and propagation of the new phenotypes in the biosphere is not the goal of the present work. It will suffice to note that there are ways of obtaining new genetic information, which have a special evolutionary significance. It can be argued that the emergence of the *fundamentally new* information in the biosphere is based on the process of mutation, which has important evolutionary implications for all single-cell organisms. At this level, there is a rapid propagation of new adaptationally significant traits owing to the extreme simplicity of the organisms themselves, as well as their high reproductive rate. The transfer of the genetic material occurs via direct interactions between the unicellular organisms, accompanied by transfer or exchange of genetic information (in the process called conjugation), also with the aid of viruses and bacteriophages, which serve as transfer vectors.

Whereas for the unicellular organisms, the acquisition and propagation of new adaptationally significant traits is a sufficiently frequent occurrence, for the multicellular organisms, this way of acquiring new traits is rather rare. Although sexual reproduction plays a significant role in the *variability* of many multicellular species by allowing random recombination of the parental traits.

Let us dwell on the last of the three conditions set out by Darwin as necessary for the evolution—*natural selection* or the driving force of the evolution. To being, we will consider the fact that various populations within the biosphere do not exist in isolation but are actively interacting. In ecology, the science concerned with such interactions, the unit encompassing all interacting organisms within a specific habitat is termed *biocenosis* or an ecological community.

In the biological tradition, the term *biocenosis* is understood as a historically formed totality of the microorganisms, plants, and animals populating a certain area of land or water mass characterized by certain interactions within itself, as well as with the abiotic factors of its surroundings. In addition, besides the "macrobiocenoses" that exist in an abiotic environment, there are also "minibiocenoses," which are most frequently observed as microflora present in virtually all members of the animal world.

For us, it is important that the existence of the various species within the biocenosis has an expressed genetic component, simply due to the fact of their coexistence in relative confinement. Here, we mean not only the previously mentioned horizontal transfers of genetic information, characteristic of the unicellular organisms that form the lower—or the basal—level in the biocenosis but also the vertical shifts occurring by the means of the same universal transfer vectors of the genetic information. These interaction and structural information transfer processes are possible because all the organisms in the biosphere "speak" the same language in the form of alternating bases of DNA, which is universal for all species. Thus, we can assert that the progression of evolution occurs as a result of the acquisition of genetic information, leaving intact the basic processes of DNA replication in all members of the biosphere.

It can be argued that biocenosis is the implementation of the mechanism of the natural selection and the fundamental unit of evolution. In other words, natural selection is aimed not at a particular species, but rather at a system of organisms interacting with their environment in a given space. In which case, the multicellular organisms are the "bridge" between the unicellular organisms and biocenosis as a whole. Let us consider in more detail the interactions between individual organisms within the biocenosis.

It is well established in biology that any close interaction of the organisms among themselves can be represented as an opposition or a predator-prey, parasite-host, or symbiosis type of relationship. The first type of interactions represents the well-known food chain, which makes possible the existence of not only the autotrophs but also all heterotrophic organisms in nature. Within this framework, the energy cycle can be traced in which the energy (most frequently solar or geothermal) and substances from the environment are continuously used to generate the living matter.

A huge number of works have been written describing the interactions between the predator and the prey, along with the accompanying evolutionary processes, which, it is important to note, are always directed at the conservation of a certain balance within the particular biocenosis. As far as the other two interaction types in the biocenosis go—parasite-host and symbiosis—the picture is quite different. The occurrence of parasitism in itself can be viewed as a victory in the fight for sustenance in which the predator itself becomes the source of food for the former prey. However, in many cases, such relationships can develop into a different class—*symbiosis*. It is quite possible that many of the evolutionary acquisitions had been obtained in such a way. The acquisition of the mitochondria by eukaryotes had its beginnings as an indigestible prey that made its way from a latent parasite to an absolutely indispensible symbiont.

Such processes of adjoining, or integration, are not only important for the existence of the biocenosis, creating it as a system, but also demonstrate *the ability of the biological systems to gain complexity*. Actually, as the first eukaryotes, the first multicellular organisms arose with the aid of the already-existent ability of the basic unicellular organism to mutually synchronize or coordinate their internal processes with each other. On the other hand, the presence of the continuous exchange of genetic information, via the viruslike particles, had created conditions for the emergence of a fundamentally new program, which worked to functionally unite a large number of cells. Such method of coexistence of cells had allowed them to improve significantly the fidelity of their genetic information. In the course of evolution, the pinnacle of such a method of organization were the multicellular organisms, an inseparable component of which is a program of their development—ontogenesis.

It could be said that the evolution is one of the means of adaptation of biocenosis in general, based on the acquisition of new genetic information. In one of the recent works, Chris Venditti, Andrew Meade, and Mark Pagel performed the computational analysis of the different scenarios by which species acquire new traits. The authors show that neither a gradual accumulation of new information nor the increased influence of the new information in the course of its acquisition correlate with the available data on the historical development of the phylogenetic tree (77). Correlation is only achieved when the model accounts for the physical (geographical) or genetic *isolation* analogous to the process of formation of a biocenotic system. It can be argued that the biocenoses emerging and disappearing over time are the constant "training ground" for the adaptation processes in the biosphere.

Returning to the phenomenon of natural selection, let us ask ourselves whether the evolutionary processes are always a response to the sudden perturbations in the properties of the environment. It is well-known that under stable conditions, the natural selection should play a stabilizing role, and under crisis conditions, a stimulating one. However, in reality, such strict correlation is not observed, but instead, a series of moments in the history of the biosphere indicates that a sudden deterioration of the climate does not always result in the "evolutionary boom." On the contrary, a rapid increase in the diversity had been observed under stable conditions. A possible explanation for this is that at a certain level of development, biocenotic systems begin to develop according to their own internal laws or reasons, which in practical terms can be described as "adaptation to adaptation" of its component organisms. The internal condition of such complex systems takes a leading role in their internal processes, which is characteristic of biological systems in general. It is appropriate here to remind the reader of a property characteristic of any supercomplex systems, frequently called the butterfly effect (even an insignificant event such as a flapping of a butterfly wing at one location can cause a hurricane at another location). Insignificant changes in a genome of one of the participants of the biocenosis can lead to a sudden change in completely different parts of biocenotic system. In such situation, a developing system can create its own critical state. The answers to the question of why some species do not change in the course of a long period of time while others appear and disappear are not outside but are contained *within* the biocenotic systems. So the evolution of the microorganisms in the direction of increased complexity does not occur not because it is impossible, not because it is unnecessary—the microorganisms are the champions of adaptivity, after all, thriving in any environment—but because it has *already happened* once a long time ago. Today, they are a part of a vast system containing organisms of different levels of complexity and functioning by its own internal reasons and rules.

The Role of Ontogenesis in Multicellular Organisms

In itself, the existence of short-lived organisms in nature presents an evolutionary mystery. As the long-lived individuals are able to produce more offspring than the short-lived ones, natural selection should favor the increase of the life span. However, the available biological data are to the contrary—the life span of the majority of species in the biosphere is relatively short, being less than a year on average (78). It turns out that rapid maturation of the organism, in combination with high reproductive rate, provides certain evolutionary advantages.

An alternative is slow maturation with a low reproductive rate, though with a significantly longer life span, possible under condition of low competition due to a unique ecological niche. This option is characteristic of the more complex organisms—slow maturation necessary for the rearing of the offspring whose further behavior is based on learning. Taking into consideration the obvious advantages of rapid maturation, it is necessary to understand why and how the long-living organisms came about and whether the biosphere contains a vector that leads to the longer life span. Once again, it is important to emphasize one of the key universal properties of the multicellular organisms related to their individual development—a *direct correlation* is observed in nature between the duration of the sexual maturation and the longevity of any multicellular organism.

Thus, the biosphere certainly has two distinct strategies or types of adaptation for evolution of the multicellular organisms. One strategy is based on the rapid changeover of the generation, which clearly reduces the evolutionary demand on the individual adaptational abilities, allowing the organisms to survive with a relatively low level of reparative abilities. The other strategy is aimed at a far slower rate of the generation changeover. Such situation becomes possible owing to the appearance of highly specialized organisms requiring a lot of time for their development. Consequently, the rate of evolutionary changes is slowed down, compensated by the high adaptability of such organisms. Achievement of such strategy allows a precipitous slowing of the rates of development as well as aging. In this case, natural selection will favor the survival of the organisms possessing advanced reparative mechanisms necessary for prolonged existence. Individual development of the multicellular organisms happens according to a distinct program generally referred to as ontogenesis. This gives us a reason to believe that a mechanism as important as ontogenesis, which implements the program of the individual development encoded by the genome, can also provide a systematic turnover of the generations within a population. Ontogenesis, as a developmental program, only appears at the level of the multicellular organisms, bringing with it age-related changes, naturally directing the organisms toward death. This is exactly the connection that provides the basis for us to begin the search for the causes of the aging process in this direction.

Let us clarify what is meant by the term *ontogenesis* in the present text. In the biological literature, the most general definition of ontogenesis for the multicellular organisms is the "individual development of the organism from fertilization (during sexual reproduction) until death." In the multicellular animals, it is divided into the embryonic (prenatal in mammals) and postembryonic (postnatal) phases. In other words, ontogenesis is understood as the implementation of the genetically encoded *developmental program*.

The main question in trying to understand the role of ontogenesis in the processes of aging is whether ontogenesis determines the processes of aging or only the development of the organism until sexual maturity is reached. Is aging simply a consequence of development, or is it directly controlled by the mechanism of ontogenesis?

The aforementioned correlation between the duration of the sexual maturation and life span speaks in favor of the first hypothesis.

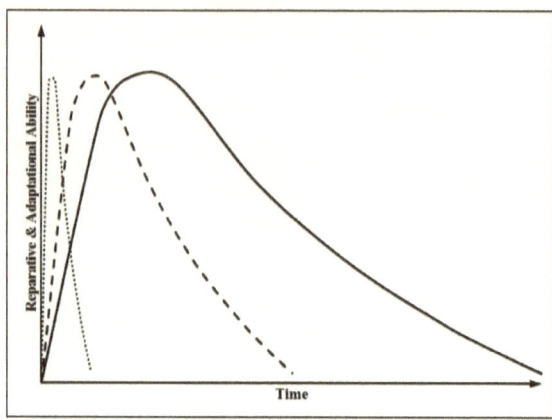

Diagram 1: Level of reparative and adaptational abilities of an organism over its life span

We get an impression that the developmental process itself provides a certain "impulse to action" for the organism, and the longer it lasts, or the stronger the "impulse," the longer the trajectory. On the other hand, we recall that no genetic mechanisms that directly regulate the life span and the processes of aging have ever been identified.

The diagram illustrating this point shows how the adaptational and reparative abilities of the multicellular organisms change in the course of their development. The shorter the rising portion of the curve, the faster the peak of development will be reached. Accordingly, the aging processes will be faster, shortening the overall life span of the individual.

The existing relationship between the life span and adaptational/ reparative resources makes it necessary to establish what defines these *adaptational resources of the organism*, and more importantly, *why* and how do the *limitations* on the consumption of hydrocarbon and protein resources that lead to the decline of the reparative abilities in all somatic cells of all organism arise.

Two Parts of the Cellular Genome

Let us examine the relationship between the organism as a whole and its constituent cells. We ask the question, what is an organism? Is it the sum of the specialized cells participating in a common process, or is it a sufficiently independent formation? From the biological knowledge, we know that the *multicellular organism* is an independent system, a complex ensemble of many cells, integrated tissues, and organs interconnected with the aid of chemical factors, which can be hormonal as well as neurotrophic. The cells of the multicellular organisms are *totipotent* (i.e., they have the genetic potential to be any cell of a given organism). Being equivalent in their genetic information, the difference between them is in the set of the genes that are expressed. This difference in genome expression leads to the variety of morphologically and functionally distinct cells in the organism and is based on the phenomenon of *differentiation*.

With the seemingly infinite variety of life-forms, there is a fairly small number of basic types of cellular structures that are the fundamental processes in the multicellular organisms. Thus, there is, in principal, no difference in the metabolism, biosynthesis, or informational exchange on the cellular level (at least in animals). It can be said that all basic structures of the cells are far more similar than they are different, so even the multicellular organisms that are far apart on the evolutionary tree of the multicellular species have similar organs and systems that function according to universal principles. This similarity is due to the structural basis of all members of the biosphere, being the amino acid sequence of the proteins and the nucleic acid sequence of the DNA, which enable a limited set of functional solutions. It is natural to assume that the same type of the structural framework imparts a certain "canalization" on the properties of the biological objects, which is expressed as the universality of their basic function on the cellular level, as well as on the level of the whole organism.

Any process occurring in the organism, including ontogenesis, is primarily determined by the genes of the cellular DNA. When a multicellular organism is considered from the standpoint of its genome, we encounter the fact *that its cellular DNA consists of two functionally distinct parts*. One part of the cellular genome contains the genes necessary for the cell to define its specialized functions characteristic of the multicellular organisms. This part of the genome, as is well-known, is present in all cells, providing them with specific functionality, depending on which genes are active in each particular case. We will denote it as *SG* (specialized part of the genome) for brevity.

The other part of the genome present in the multicellular organisms is known as *the housekeeping genes* (79-83). We will denote it as *HG*. This is the part that contains the genes that ensure the existence of each individual cell. It can be called the universal platform on which evolution and ontogenesis is based in the multicellular organisms. Such division of the genome is justified not only from the standpoint of its function but also from the point of view of the evolutionary origin of multicellularity.

Thus, the monograph of A. Boldachev contains the following contemplations:

> *In its essence, original to nature, the genome (genetic mechanism) is responsible for the function of the individual cell, and that is all* (!): on the basis of the DNA fragments, the enzymes (proteins) are synthesized, which regulate all processes of the internal and external life of the cell. Genetic mechanisms of the specialized cells of multicellular organism do not contribute anything fundamentally new to these processes. The only significant difference of the specialized cells is their *limited use of the entire genome of the organism, fixed for the duration of their whole life.* (Eukaryotes have possessed the capacity for temporary, reversible morphological modifications long before the formation of the multicellular organisms.)
>
> These rudimentary considerations lead to a simple idea: a *genetic mechanism, in the course of supporting the life of the cells in a multicellular organism, does not have a direct*

relationship to the functioning of the organism as a whole. The genome does not contain functional mechanisms of regulation on the level of a system of cells. For the genome, the organism does not, in principle, exist. (84)

We can agree with the author in that, at the cellular level, the genes of the organism itself do not participate in the processes of the cell life, leaving this role for another part of the cellular genome. The term *minimal genome* is used in the works dedicated to the creation of an artificial genome based on the mycoplasmal DNA (85) and is very close in its definition to the housekeeping genes. In fact, the goal of the reproduction of the simplest possible genome of a bacteria itself has led to the creation of this term. By definition, such a genome must contain only the genes essential for existence of any organism starting with the simplest one. Thus, any biological system requires a minimal or *an invariant* set of structures and processes that support the existence of the system itself. Let us imagine what such a set, universal for any system, would include.

First, there are the processes that preserve the thermodynamic equilibrium of the system (constancy or the reduction of the entropy of the system) in the course of the energo-material exchange with the environment. The energo-material exchange occurs because of the existence of the relevant structures in the system.

Second, there is the process to support continuous *replication* of the structural contents (method of organization) of the system itself. This process is realized via the storage of the information about the structure of the system with the aid of a kind of a structural template or *the genome*, which enables its exact replication. Moreover, this is where the ongoing repairs of the genomic structure itself, damaged for one reason or another, happen. In the living organisms, this process essential to life is called the DNA *repair*. In other words, the presence of the genome enables the process of formation of the structures that are, in turn, necessary for the basic processes occurring in the system. It also defines the specific method of the cellular organization.

In addition, we need to point out the processes that provide the *integrity* of the system, its existence in conjunction with all the interactions occurring within it. These are the processes that ensure the intrasystemic synchronization and coordination and storage of the pertinent information, as well as functional coordination supported by the downstream and upstream connections. It is these processes, in the end, that give rise to a so-called autonomous or *operationally closed* system. Accordingly, the genes that ensure the functioning of all present invariant processes compose the minimal genome of any cell. It can be concluded with certainty that it is exactly the activity of the genes that make up the minimal cellular genome, or *HG*, that opposes cellular aging.

Principles of Cell Interactions with the Environment

Any cell or biological system exists in an environment and continuously interacts with it. Attainment of its main objective—self-preservation—is addressed by the adaptational reactions. It is known that any biological object strives to maintain its equilibrium with respect to the environment. In biology, this condition is termed *homeostasis*—something that is necessary for the existence of the system, the "range of values" of its key parameters. From the biological point of view, adaptation is the processes directed at preservation and maintenance of the homeostasis in the system. This means maintaining it within the limits of invariability, in other words, within a rather broad range. In other words, the state of homeostasis is a "moving target" rather than a steady state.

As for the concept of *adaptation*, it is not as clear-cut. Essentially, the term *adaptation* usually refers to the processes in the system that are directed at the preservation of the homeostasis. However, this does not at all mean that the adaptational processes themselves cannot move the values of the key parameters of the system from the *ideal* values. On the contrary, adaptational processes in the system lead to continuous fluctuation of its key parameter, especially when a property of regulation such as adaptation based on experience, characteristic of complex biological systems, is implemented. We will try to consider these processes in more detail.

Any adaptational responses in the system are based on the initial basic description of the system itself. Let us clarify this assertion. Any response of the system occurs only when a difference arises between the current and the system-set value of a parameter being regulated. The presence of such preset values—the "gold standards"—or the range of response gives the system an ability to adapt to the current situation. Here, not only the quantitative aspect is important but also *which* parameters are included into this list of standards. In this way, the biological systems cannot directly react to a range of external influences. Take ionizing radiation as an example. Its presence is not sensed by the cells directly but is perceived just as any other damaging influence requiring the activation of the DNA repair processes. It can be argued that the existing information in the cell completely determines how it will respond to which influences during its interaction with the environment. In addition, the environment is not directly able to alter the response mechanisms that exist in the cell. The only possibility for such change is the acquisition of the conceptually new information in the course of evolution.

Adaptational responses connect any biological system with the environment only at the level of the energo-material exchange while following the laws of thermodynamics. Such processes are well described within the framework of thermodynamic or binary approaches and models. However, we must reiterate here that the *external environment* is always the object of consumption by the biological system. For any biological object or system any external, relative to itself, surrounding is a source of its hydrocarbon and protein consumption for which it is intended. At the same time, the external environment produces a constant, nonspecific external influence on its internal parameters. For a biological system, the response to the external stimuli is determined by its own makeup, represented by its genome and, correspondingly, not having a direct relationship to the external environment, leading to the situation of a closed system. Accordingly, the biological systems do not produce any operational, or binary, effects on the external environment. The internal regulatory processes of the biological systems are completely indifferent to what is happening in the external environment. It is important to note that the regulatory mechanisms on the cellular and organismal levels were not created for intentional interactions with the external environment, do not include it in the range of the downstream and upstream interactions, which makes the events of the external environment causally independent. A detailed description of the properties of such operationally closed systems can be found in the works dedicated to the theory of apoptosis (86).

Returning to the known phenomenon of age-related decline of the adaptational abilities, we can argue that there are no direct causes for this phenomenon present on the cellular level. With exception of an extremely rare case, the external environment does not make excessive demands on the cells, and an indefinite existence of such biological system does not contradict the laws of thermodynamics. It can be asserted that the *HG* part of the cellular genome *does not contain* the causes for the appearance of the aging processes.

The *adaptational and reparative resources* of an individual cell (or a basic biological system) are directly determined by the power of the biosynthesis controlled by the *HG* part of the genome.

Relationship between the Organism and Its Cells

It is generally accepted in biology that any processes occurring in the organism can be viewed on the cellular or the organismal level. These levels reflect the structure of the multicellular organisms, as well as the functional subdivisions of their cell's genomes. Thus, the cellular level of the organism function is determined by the work of *HG* part of the genome and the organismal level by the *SG* part. The separation of the genome into two parts raises a question of how these two levels of organization interact and how does the competition for the resources in the cell arise in the course of ontogenesis and, most importantly, in the course of the *aging processes*.

It can be argued that on the cellular level, or on the level of the basic systems that make up the organism, there are no fundamental restrictions that limit their life span. A conclusion follows in agreement with the existing biological data that the phenomenon of aging itself is a property characteristic only of the multicellular organism.

Let us try to understand whether the organizational method of the cells in the organism itself can give rise to the processes of aging. To this aim, let us consider what the multicellular organism is from the standpoint of implementation of the cellular genome contained within *SG*: How did the multicellular organisms become possible, and how do they function? How is the problem of self-organization solved under these conditions?

As was noted earlier, in order to survive, any biological system requires an invariant set of processes along with their supporting structures. In order for a multicellular organism to exist as a *self-sufficient* biological system, it requires at least the same invariant base set as is present in its component cells. It is important to emphasize here that any biological system can only exist independently as an ensemble under the condition that *all* its component parts functionally interact. Really, independent existence of the components of such a system would be meaningless, just as the separate parts of the engine are meaningless when it is disassembled. Only when it is assembled and supplied with fuel can it be started.

That is exactly what happens in the course of development, or embryogenesis, in multicellular organisms. This phenomenon can be observed in the course of ontogenesis (its prenatal phase), which, as we know, mirrors phylogenesis—from a unicellular zygote develops a multicellular organism. The cells, which are all of the same type in the beginning of the organism's development, acquire various abilities or properties necessary for the existence of the organism during the process of *differentiation*. Later, we will return to this phenomenon. At this point, it is sufficient to emphasize that the diversity of the species in the biosphere is due to one major factor—the phenomenon of specialization or cell differentiation. These processes, supported by the function of *SG* part of the cellular genome, are the foundation on which the evolution of the multicellular organisms is built in nature.

To briefly summarize, we can assert that *specialization* is the separation of the main objectives, or the basic processes themselves, which must be carried out by any primary biological system (unicellular organism), into their components. In this way, the *organismal functions* of the multicellular organism arise in individual cells, tissues, and organs, and the adaptational responses and processes transpire as they do in the members of the primary level of organization—the unicellular organisms. An organism as an independent system becomes possible when, with the aid of the allocated functions, the previously outlined set of the fundamental processes is reproduced. A situation arises where the multicellular organism acquires a series of novel abilities, primarily due to practically unlimited range of adaptational responses.

What advantages does a multicellular organism acquire? How does it "pay" for them? Note the fundamentally important point: diversion of a part of the hydrocarbon and protein resources of the cells of an organism, which function as its component parts, reduces their individual adaptational abilities—functioning of a part is always parasitic relative to the whole. However, as a result, the adaptational abilities of the organism as a whole increase, which gives an overall advantage to its separate functional elements. This happens due to the reduction in the demands placed on the cells by the external environment, which, in this case, are largely determined by the processes in the system as a whole. From this, we can conclude that the phenomenon of the multicellularity does not fundamentally increase the level of adaptational abilities but is rather *an alternative path toward higher complexity* for a biological system as a whole.

Multicellular organisms can be considered as a product of the implementation of a certain "viral program" developed in the course of evolution and executed in the cells as the *SG* part of the genome and directed at a significant *increase in the preservation of the genome as a whole*. At the same time as the appearance of the cellular functions occurring in the course of ontogenesis, the program for the development of the multicellular organism as a self-sufficient *functional system* is also fulfilled. We emphasize that this is specifically a functional system, in other words a self-sufficient subject, and not simply a sum of functions.

The potential abilities of such a system (an organism) with respect to the diversity of the functions carried out within it are immense, as is clearly observed in the example of wildlife. It is worth noting that in the case of the evolutionary increase in complexity within a functional system, the fundamental, invariant functions begin to partition into the ever-increasing number of subfunctions, creating certain autonomous subsystems and their corresponding structures—organs and tissues—in a living organism. Significant increase in complexity forces the functional system to build up a matching architecture for its regulation based on the significant autonomy of the individual subsystems and a certain "hierarchy" in the course of the evolutionary processes. In this case, the leading role is given over to the regulatory processes aimed at coordination or, as one could say, processes of "adaptation to adaptation." Coordinated functioning of the system requires the presence in it of a huge number and a particular architecture of the horizontal and vertical information conduits that carry out the coordinating function. The presence of distinct higher regulatory centers in the multicellular animals does not at all mean that these centers govern all functions in the organism. Their function is to preserve the fundamental, genetically determined parameters of the system as a whole within the range of acceptable values. Such regulation is attained via the coordination of the adaptational processes at the lower hierarchic levels. The aptitude of the biological systems to adapt, based on the preservation of the experiential information, takes on an immense role in the case of the multicellular organisms possessing a developed nervous system.

This type of regulation cannot be fully described with the aid of the binary analysis of the system. An alternative and more flexible description is necessary for the analysis of these processes. In conclusion, note that the study of the processes of development of the individual hierarchical levels and their corresponding specific adaptational processes in the complex systems is a distinct field with great prospects deserving of attention.

Returning to the main problem addressed in this chapter, let us try to isolate the most important question: What new properties different from the unicellular organisms are acquired by the multicellular functional system? The main paradox present in such system (multicellular organism) is contained in the inability of the system to directly "turn on" its constituent elements, its cells, due to its own operational partitioning, a fundamental inability to interact directly with a cell or an element, which for the organism as a whole, is a part of the external environment.

Both levels—the cells by themselves, where the processes are regulated based on the operation of the *housekeeping genes* (*HG*), as well as the organism as a self-sufficient system controlled by the operation of *SG*—are inverted with respect to each other, becoming a part of the *mutually external environment*. An operationally closed system cannot maintain its separateness when it is completely absorbed by any other system without losing its causal circularity, the status of distinctiveness. It turns out that for the primary systems (or, as in the case of the multicellular organisms, its component cells), the external environment is the cellular microenvironment composed of the entire organism and, in some cases (such as the existence of the tissues bordering with the external environment), the external environment itself. On the other hand, an organism as a system performs its required processes inside its own "space-time continuum of regulation," using the abilities provided by its elements, the cells. In this case, interaction of the organism with its cells occurs without direct interaction with them; it simply "consumes" them as an inseparable part of its external environment. We emphasize that both of these levels of organization—cellular and organismal—are physically situated in the same cell of the organism and are represented by the previously discussed parts of the genome (*HG* and *SG*).

Such situation is reminiscent of the main paradox in quantum physics where the basic particle is simultaneously a particle and a wave. Indeed, in such complex systems that include multicellular organisms, their basic component elements, the cells, simultaneously act in two capacities—as a self-sufficient system and as a part of a larger system, upholding the prediction made by Niels Bohr with regard to the paradoxes in biology.

As an example, we can use this well-known image in which anyone can see either a vase or two faces—depending on the "tuning" of his or her internal point of view. In much the same way, we can focus on either seeing the organism as sum total of its cells or as a self-sufficient system consisting of cells.

Any operationally closed system "describes" or "imagines" its external environment based solely on its own construction. Contents, or construction, of the system determines its methods of interaction with its surroundings. In this case, the meaning of any processes in living systems, as was previously noted, is simply to exist. Here, it is necessary to return to the main problem to which this work is dedicated: What gives rise to the adaptational abilities of the organism and counteracts the aging processes? As we noted earlier, they are based on the activity level of the *HG* part of the cellular genome. However, it remains unclear what limits its activity, *where* and *why* does the competition for resources occur? These are the questions addressed in the next section.

The Role of Dividing and Postmitotic Cells in the Aging Processes

We are approaching the resolution of the main question, namely, *how*? Without the turnover of the generations, the evolution, in principle, is impossible. However, so far, it is unclear how evolutionary, ontogenetic, and aging mechanisms work and what they are based on. Following the conclusions reached in the sections above, we can offer a description of this mechanism.

One of the main suppositions, on which the current work is based, is the functional division of the cellular genome into two parts—*HG* and *SG*. It is from this standpoint that the further analysis is constructed. Let us consider how a multicellular organism develops from the viewpoint of the operation of the two previously described parts of the genome. In order to do this, it is necessary to trace the intensity of operation of *HG* and *SG* parts of the genome, as well as the roles of the two types of cells that the organism is composed of—*dividing* and *postmitotic*—in the course of this process.

First, it should be noted that the performance of its specific function by the cell and the process of cell division are incompatible as the period in the cell cycle when the synthesis of the specific proteins is possible is rather short. The constant on and off cycling of the function cannot support the normal operation of the cells, which compose the functional units of any organism. In the multicellular organisms, this obstacle is overcome via existence of two separate cell types—dividing cells and postmitotic cells—incapable of further division.

In turn, two types are always present within the population of the dividing cells: stem cells and differentiated cells. Moreover, it is the differentiated cells that enable continuous performance of a specific function of a particular tissue or organ. The role of stem cells is to maintain the constant quantitative composition of tissues or organs by compensating for the continuous attrition through dying of the terminally differentiated cells. These are the cells that require the telomere clock, which limits the number of divisions a specialized cell can undergo. It is worth noting that the division of a stem cell always results in production of a new stem cell and a cell with a strictly determined differentiation. As a consequence of these processes in all tissues and organs of an individual composed of *dividing* cells, the function they perform is continuously maintained at an optimal level. This level of function that is taken in an abstract tissue and is considered separately from the organism can be preserved indefinitely.

Interestingly, the various relationships between the intensity of function of *HG* or *SG* parts of the cellular genome correspond to the various states of the cells. If we look at the ratio of the expression levels of *SG* or *HG* parts per unit time in the cell, then for the terminally differentiated and postmitotic cells, this ratio *HG/SG* is always less than 1. For stem, embryonic, and tumor cells, this ratio is always greater than 1.

However, it should be remembered that all organs and tissues consisting of continuously dividing and potentially dividing cells are part of a unified system (an organism), although these cells are only one perpetually turned-over part of an organism representing all the epithelium, bone marrow, and in some cases connective tissues. In this context, it seems unlikely that the Hayflick limit (the final length of the telomere, which determines the limit on the number of cell divisions according to the telomere clock hypothesis) plays a decisive role in the aging processes. The perpetually dividing cells and tissues formed by them play only a role of a continually replaced, disposable tool.

Other cell types in the organism are highly specialized. These are the so-called *postmitotic* cells. The cells of this type make up all the vital organs, which have a complex architectural structure reflective of their function. Following formation, such organs have a severely limited ability to repair themselves. It is practically impossible to regenerate lost parts of the skeleton, muscle, kidneys, lungs, myocardium, and especially nervous tissue, precisely because of their complex architecture. It is natural to say that the decisive influence on the condition and the adaptational abilities of an organism and, consequently on its age-related processes, is exerted exactly by the postmitotic cells. Formation and growth of all complex tissues and organs are completed just as the organism reaches sexual maturity, which integrates the multicellular organism into the chain of transmission of hereditary information. It is in this period of sexual maturation, when the implementation of the program of development encoded in the genome completes, or the postnatal period of ontogenesis, that the processes of age-related changes begin to arise. Let us look at what processes associated with the expression of the cellular genome take place during this time.

Cellular Membranes and Processes of Aging

At present time, it is known that all structural and functional proteins that perform a specialized role are synthesized in the cell, on the polysomes associated with the membrane, or on the endoplasmic reticulum (ER). At the same time, proteins necessary for the function of the cell itself are synthesized on free polysomes (87). An exception from this are the membrane proteins, which make up the structural and functional basis for the membrane-bound complexes in the cell.

Specific structures of neurons, myocytes (muscle cells), muscle fibers, hepatocyte oxidases, and all secreted proteins, as well as specific proteins in all epithelial tissues, are either associated with, or are parts of, membrane structures. In other words, we can assert that all proteins encoded by the *SG* part of the genome are synthesized on membrane-bound polysomes, whereas *HG* proteins, with the exception of the membrane proteins themselves, are synthesized in the cell by free polysomes.

Such situation is biologically justified. After all, it is only natural to imagine that the location of synthesis and the location of function are interrelated. It has been experimentally determined that the ratio between the number of free and bound polysomes changes to favor the latter, while in malignant growth, for example, the number of the free polysomes increase. We can conclude that the ratio of free and bound polysomes reflects the ratio of expression of *HG* (free polysomes) and *SG* parts (bound polysomes) of the cellular genome.

Mechanisms exist in the cell to process the proteins that are synthesized by it. In particular, a mechanism directing proteins to either free or bound polysomes based on the presence of a lipophilic region in the beginning of the polypeptide chain. Because of this region, the nascent protein with its polysome attaches to the external cellular membrane or to the endoplasmic reticulum (ER). During the posttranslational processing of the polypeptide, the lipophilic region is cleaved off, and the nascent protein either remains attached to the membrane, is transported through it or, in some cases, is inserted into a membrane-enclosed compartment.

At present, no evidence exists for difference in the rate of transcription of RNA of *HG*—and *SG*-encoded proteins. Similarly, such evidence does not exist for the difference in the rates of translation. However, there is sufficient experimental evidence suggesting that at the level of transcription and translation, the rate of operation of the *HG* part of the genome is regulated primarily by a feedback mechanism inherited by the cell from its unicellular ancestors and of the *SG* part by a feedback mechanism at the level of the functional unit of an organ or at the level of the neurohumoral regulation of the organism as a whole.

As was previously noted, the ontogenetic program of the organism must pursue two major goals—to achieve the ability of an individual to reproduce and the generational turnover necessary for the evolution of the species. In this regard, the central part of ontogeny is sexual maturation, a time when an individual fulfills its, evolutionary in significance, role as a species. Along with cessation of growth of the organism (increase in linear dimensions and mass) a period of sexual maturity begins—maturity of all its functional systems supported by the function of the *SG* part of the genome.

It can be concluded that the rate of development of an individual is determined by the degree of "switching" of the expression levels of the *HG* part of the genome in the cells to the *SG* part, which is necessary for the formation of all functional organs and tissues. This situation accurately reflects the law of ontogenesis according to which the rate of growth and development is strictly proportional to the rate of degradation. In other words, the slower the growth, the greater the life span. The point of the maximum adaptational stability of an organism occurs during the period of sexual maturity and corresponds approximately to the point of the cessation of growth. Specific, genetically determined, linear dimensions of an organism are attained by a certain number of cell divisions, the limit of which is set by the terminal differentiation of the cell with the subsequent loss of ability to divide, which leads to a gradual increase in expression of *SG* part of the genome. The final formation of the physical dimensions of the organism is completed by the hormonal system, as it happens in mammals, for instance, where the somatotropic hormone (STH) is involved in setting the size of the organism by regulating the growth of the long tubular bones. We should add that at one time, the hormonal system was considered to have one of the leading roles in the processes of age-related changes. However, while the hormonal system can accelerate or slow down the processes of aging, its significance does not go beyond the role of transmitting the genetic program of development present in the organism (72, 88-90).

The early part of ontogenesis is clearly dominated by the expression of the *HG* part of the genome, which provides the necessary growth and development of the individual during this time. In other words, in the course of the first half of the postnatal ontogenesis, a clear advantage is observed of the "production of the means of production" over the "production of the consumables." With growth and development of the organism, its adaptational abilities increase, reaching their maximum value by the period of sexual maturity.

We can illustrate the course of these processes with a diagram.

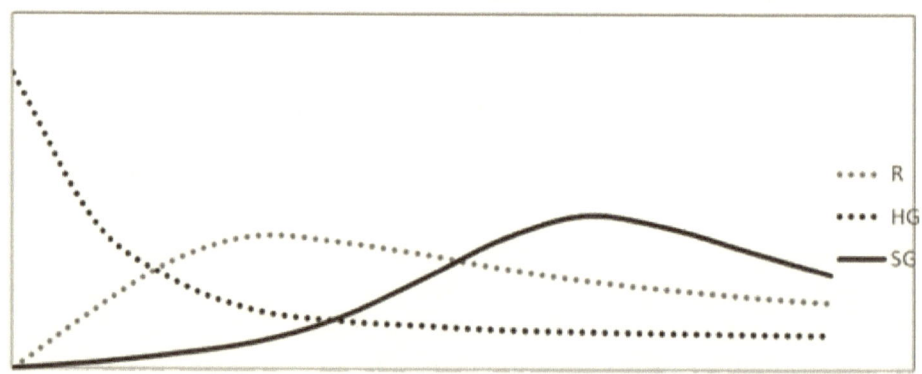

Diagram 2: Level of HG and SG activity and
reparative abilities of an organism over its life span

The x-axis represents time, and the y-axis is the level of expression of the two parts of the genome—*HG* and *SG*. The horizontal dashed line signifies the minimal level of adaptational and reparative abilities. The first curve shows the level of production of *HG* and the second of *SG*. The solid line represents the resulting curve—adaptational and reparative abilities of the organism in the course of a lifetime.

As follows from the illustration bellow, in the course of development, an organism passes the peak of its abilities in the range of cessation of growth and reaching of sexual maturity. Further increase in the portion of expression of *SG* invariably leads to decline of the synthesis of the products of *HG* and, as a result, the lowering of the adaptational and reparative abilities of an organism.

As we can see, the basic properties of the multicellular organisms, namely their adaptational and reparative abilities, are determined by the expression ratio of *HG* part of the cellular genome of the cells themselves and *SG* part responsible for the functions performed by these cells, which make up the organism with all its functions as a whole. It is here at the level of expression of the two parts of the genome that the *competition for resources* arises in the cell.

Recall that for an organism as a whole, the decisive role in these processes is played by the postmitotic cells, which constitute the basis of a multicellular organism. These cells determine the level of adaptational abilities and available resources in the organism, which is in turn dependent on which functional part of the genome (*HG* or *SG*) is prioritized. This prioritization is based on the total surface area of ER, which diverts cellular resources and, more importantly, reparative abilities to itself.

What is the *cause* of the increase of the ER surface area in the course of aging? The first supposition is that, following the cessation of cell division, the surface area of intracellular membranes tends to increase by inertia. An objection can be made that the preceding growth of the ER membranes is not necessary to the cell as the increase of their surface area is necessary not *before* but *after* the process of cell division. It is more likely that such process is directly linked with the expression of the *SG* part of genome. We can hypothesize that there is a specific mechanism for the preceding ER growth in the nondividing cells presented in the *SG* part of the genome. Such hypothesis is only logical, as such mechanism is necessary for the *SG* part to perform its function. However, in the opinion of the author, it is more likely that the *SG* part simply has the ability to stimulate, or *control*, the surface area of ER as needed by employing the *HG* part of the cellular genome for this purpose.

Still, if the control of the rate of ER synthesis occurs in the specialized cells according to the principles of feedback mechanisms, then *why does the level of this synthesis not stop at the equilibrium point but continues to grow with the consequences we have already learned?*

That is the central question essential for the analysis of the causes of the aging mechanisms. Recall that one of the basic properties of any open thermodynamic system is the constant fluctuation, a delicate balance of their parameters. As was noted earlier, due to the huge number of connections in such system, we can talk about a continuous adaptation to adaptational responses occurring in the cell itself. Such state, even without significant external stresses, continuously stimulates all functions of the organism as a system, creating a situation of constant demand for the strengthening of the adaptational response. At the same time, as was previously shown, the internal needs of the cell are controlled only by its own internal connections. Processes at the organismal level sense their cellular basis as just a part of the external environment without including it in their regulatory system. The result is that all processes of the replication of the synthetic machinery of the cell and all reparative processes are inconsequential or invisible, for the multicellular organism as whole as a higher-level system. Importantly, this situation becomes fatal *only* in the case when it arises within a space *limited* by the physical dimensions of the cell. This is exactly the situation that occurs upon the completion of the growth of the organism and the increase of the size of the postmitotic cells.

Again, after reaching sexual maturity, a gradual increase of the intensity of the load on the *SG* part of the cellular genome is observed, leading to a steady decline of the intensity of the *HG* part. In this way, reciprocal, competitive relationship between the expression of *HG* and *SG* parts of the genome of the differentiated cells gradually shifts toward the expression of the *SG* part. This situation is also associated with the limitation of the physical growth of the cells themselves. In this case, the reduction of the adaptational and reparative potentials of the cell occurs as a result of gradually increasing portion of the expended hydrocarbon and protein resources being diverted to synthesis of membrane complexes and the functional products associated with the products of the *SG* part, as well as the concurrent decrease of the volume of the available cytoplasm and free polysomes where the synthesis of the structures encoded by the *HG* part occurs. This process is self-accelerating in nature as the number of attachment spaces for bound polysomes, synthesizing proteins encoded by the *SG* part of the genome, increases as a function of the increase in the surface area of ER.

The proposed analysis explains a series of theories of aging, linking this phenomenon with age-related damage of the genetic apparatus of the cells of an organism. What we are talking about is a constant increase of the free radicals and peroxides, concomitant to the process of cellular respiration, with age. It is these compounds that are the cause of DNA damage of the cellular genome. Reduction of the metabolic rate and even appearance of wrinkles with age is associated with them. As is known, processes of cellular respiration are performed by the mitochondria—cellular organelles that are relatively independent from the cell itself. However, a gradual increase of demands from the organismal functions described earlier leads to the overloading of the mitochondria, causing their disintegration (apoptosis) accompanied by the release of a large quantity of incompletely oxidized metabolites and free radicals. The process is even more damaging to the DNA and the enzymes of the cell, considering the resulting lack of production of the reparative and antioxidant proteins.

There is another group of theories of aging that consider disruptions of the central regulatory processes of the organism as its main cause. Reduction of adaptational and reparative potentials in postmitotic cells occurs concurrently in different tissues and organs; however, the greatest role in the process of aging must belong to the nervous tissue. Among the cells making up the nervous tissue, there are both the cells with the potential for division and postmitotic cells; moreover, it is due to the latter ones that the function of the nervous tissue is carried out—unification of various tissues and organs into a single whole and regulation of the organism based on the data received from the external environment and on data "recorded" in the nervous system, both acquired and inherited experience. The spatial structure of the neurons itself, corresponding to their main function, makes it impossible for such cells to divide. Thanks to their position at the top of the control hierarchy of the organism (meaning, the neurons that make up the higher vegetative centers of the brain), any disruption of their function caused by the reduction of the adaptational abilities of the cells, in the opinion of the authors who follow the neuroendocrine theory, should play the leading role in the triggering of the mechanism of aging in an individual. Indeed, it is impossible to deny the role of the higher vegetative centers in the processes of aging; however, recall that the presence of defined higher regulatory centers in multicellular animals does not mean that it is them that govern all functions of the organism. Their function is to preserve the basic, total parameters of the system as a whole within the range of allowed values by coordinating adaptational processes of the lower hierarchical levels. It can be argued that any disruptions of the higher vegetative function regulation would accelerate the age-related change, overloading and not protecting all the specialized tissues of the organism. Although the main cause of age-related changes is still other processes discussed earlier in this work.

Concluding this section, we return to the phenomenon of ontogenesis itself. Based on everything previously said with regard to the age-related processes, we can draw the key conclusion: ontogenesis determines *only* the rate of formation and growth of an individual, leaving the natural degradation and death of an individual to the *method of organization*, or the principles of construction of any multicellular organism. It can be said that in the course of development and growth, an organism creates the very "magic skin," which is the source of its existence and which, in the course of its further life, it uses up.

Conclusions and Perspectives

Let us present the main conclusions that follow from the proposed view of the mechanisms of aging. First, we note that the specific universal mechanism of aging was not, in the end, found simply because it does not exist. However, there exist features of the basic construction of the phenomenon of multicellularity itself. These properties, which can be roughly reduced to the absence of feedback between an individual cell and the organism as a whole, are the cause of all manifestations of the phenomenon of aging.

Within the framework of the proposed approach, all main phenomena occurring in the aging organism are explained. It becomes clear that the existing ratio between the duration of growth and the life span of an organism is based on that, following the cessation of growth (completion of the ontogenetic program), the production of the *SG* part of the genome begins to dominate in all postmitotic cells at the rate proportional to the rate of the growth of an organism. As a result, the physically limited amount of hydrocarbon and protein resources of these cells is gradually shifted from the reparative abilities to the execution of external functions. This explains the increased amount of damage to the genetic apparatus of the cells in an organism by free radicals.

The disruption of the neuroendocrine regulation with age is also explained. By virtue of their functional role in the organism, the neurons of the higher vegetative centers become the bottleneck, the disruptions of which begin to break down the function of coordination and harmonization of the vital organismal processes they perform.

The proposed description of the phenomenon of aging in the multicellular organisms does not only satisfy the criterion of novelty but also of verifiability. The views presented are reasonably easy to confirm experimentally, opening a new direction toward detailed understanding of not only the processes of aging but also specific mechanisms of implementation of the program of ontogenesis in multicellular organisms. However, in the opinion of the author, the most important result is the new prospects of real impact on life expectancy. Let us try to imagine what possibilities and limitations exist in this regard.

If any organism can be defined as a "wrapper for the sex glands," then, in the case of a human, this "wrapper" is also able to comprehend its own existence. It is exactly this self-awareness that forces the humankind to think about the indispensible property of its body to age and die. From the human perspective, the inevitability of aging does not at all mean the necessity of this phenomenon, continually pushing them try to discover an elixir of youth. Based on the views presented in this work, let us try to answer the question, Is there a real possibility for the creation of such an elixir? In order to answer this question, we will begin by considering the fundamental limitations that arise on the path to the creation of an eternal organism.

The phase transition from unicellular to multicellular organization successfully, from the standpoint of evolution, combines the ability to develop with the natural degradation. However, it should be noted that such situation only arises under conditions of cessation of cell division or terminal differentiation. In plants, continuous division and regeneration of their parts, which can be observed in trees, leads to an apparent absence of limit to their life span. In this case, however, we are simply dealing with the implementation of the developmental program based on the continuous replacement of *all* structures of a multicellular organism. As a result, only the external shape is preserved, the organizational method of the given organism, which is continuously filled with new elements.

A similar phenomenon occurs in a number of representatives of the animal kingdom, for example, in the jellyfish *Turritopsis nutricula* (91). In this case, the reaching of sexual maturity triggers their program of ontogenesis from the start, renewing all their cell structures. Another example are the well-known, long-living tortoises of the Galápagos, whose growth does not stop upon reaching sexual maturity, can be taken as another example. As a result, representatives of this species die not because of the aging processes but due to the loss of effectiveness of their organism due to its physical dimension. Indeed, perpetuation of growth allows an organism to maintain exceeding level of activity of the *HG* part of the genome, compensating for the increasing demands of *SG*. However, this situation cannot continue for a long time—continuous increase of the size of an organism leads to the reduction of its adaptational abilities. It seems that this strategy of resistance to aging is also unlikely to be considered promising.

The experimental results obtained in the course of studying the effects of growth hormone on the life expectancy (88-90) indicate that even virtually complete blockage of its effect on the cells of an organism is only able to slightly increase the length of the growth phase of an organism (limitation of the caloric intake has a similar effect [90]). Only a portion of the ontogenetic program of development of an organism is accomplished with the aid of the hormonal system, and this program cannot be stopped by the hormonal system alone.

The bottleneck that determines the life span of the multicellular organisms are the postmitotic cells and the tissues they make up. This is especially true of neurons that compose the central nervous system of the organism. The neurons play the role of the substrate for the signaling function in the organism—sensing, *storing*, and transmitting information significant for the organism (commands and signals). These cells make up the stable network of informational connections, pervading the entire organism, connecting all its parts.

Any organism containing a nervous system consisting of nondividing neurons within its structure is prone to acquisition of its own unique identity. Such situation naturally leads to a fundamental limitation of any complex systems or, in this case, organisms. This fundamental limitation is due to the fact that in such closed system, there is a constant increase in the amount of the information stored in its memory. The ability of an organism to store the received information not only continuously increases the cost of its storage and use but also imposes a physical limit on the storage capacity based on its physical size. The passing of time itself, in a closed system, can only be possible with the accumulation of irreversible structural additions (91). In addition, the method of the structural organization itself, used by memory based on simultaneous function of a huge number of neurons, does not allow for the ability to "delete" old information; it is impossible to exclude old information without damaging the entire system of its recognition and storage.

If we consider the problem of preservation of the identity of an organism, we can assert that any prolonged preservation of such requires continuous increase of the capacity of the individual's memory, which is only possible to a certain limit. An alternative to continuous growth of an organism is an attempt to stop the internal time within the organism itself by stopping the increase of expression level of *SG* part of the genome. However, such path will naturally lead to the appearance of an organism whose "volume" of memory and uniqueness will be strictly limited. It is exactly in these limits, the values of which are not yet known, that it is possible to prolong the life of an organism while preserving its unique identity.

The author agrees with the proposition that in the future, the more and more elaborate attempts to prolong the individual life of the body will not be the most promising. The main path, along which the human scientific investigations will develop, is the creation of direct connectivity of the human mind to digital devices (92-93). The emergence of such ability will make possible the preservation of an individual consciousness for a time simply incomparable with the life span of a biologically based organism. However, this is a distant possibility; while at present, in the opinion of the author, it is worth focusing on the possibilities enabled by the approach to the understanding and control of the mechanisms of aging offered in this work.

Given the relatively narrow and specific scope of the problems addressed by the present work, quite a few "branches" have been left out of it. If within the scope of this work, the relationships between the organism and the cell, or the "systems of systems," were considered, the analysis of the systems of the next level is also of interest, designed to point out specificities of the processes occurring in biocenoses and possibly in social communities. An interesting problem of the regulation in complex systems exists, where "strategic algorithms" start participating in these processes directed at altering the systems future. It is entirely possible that such systems will become the most important in the future.

In conclusion of the present work, it is necessary to briefly outline the possible ways of its practical implementation. It is clear that the next step must be the experimental verification of the presented proposals and views of the author with regard to the mechanisms of aging. Introducing the prospects being opened by the new look at the basis of the mechanism of aging, it can be argued that the result of their practical implementation may well be a considerable extension of the active phase of human life. To this goal, it is necessary not only to study in detail the mechanisms of action of the *SG* part of the genome on the growth of the endoplasmic reticulum membranes but also to understand how manipulations of this mechanism affect the storage of information and its reproduction in the neurons of the nervous system. It is worth noting that even a result allowing to extend the life span of a human one and a half or twofold cannot fail to inspire the efforts to experimentally test and practically implement the offered view on the mechanisms of aging.

Notes

1. Szilard, L. (1959). "On the nature of the aging process." *Proc Natl Acad Sci USA* **45**(1):30-45.
2. Orgel, L. E. (1963). "The maintenance of the accuracy of protein synthesis and its relevance to ageing." *Proc Natl Acad Sci USA* **49**:517-521.
3. Beckman, K. B., and Ames, B. N. (1998). "The free radical theory of aging matures." *Physiol Rev* **78**(2):547-581.
4. Hamilton, M. L., Van Remmen, H., Drake, J. A., Yang, H., Guo, Z. M., Kewitt, K., Walter, C. A., and Richardson, A. (2001). "Does oxidative damage to DNA increase with age?" *Proc Natl Acad Sci USA* **98**(18)):10469-10474.
5. Ames, B. N., Shigenaga, M. K., and Hagen, T. M. (1993). "Oxidants, antioxidants, and the degenerative diseases of aging." *Proc Natl Acad Sci USA* **90**(17):7915-7922.
6. Barja, G. (2002). "Endogenous oxidative stress: relationship to aging, longevity and caloric restriction." *Ageing Res Rev* **1**(3):397-411.
7. Comfort, A. (1964). *Ageing: The Biology of Senescence*. Routledge & Kegan Paul, London.
8. Cristofalo, V. J., Volker, C., Francis, M. K., and Tresini, M. (1998a). "Age-dependent modifications of gene expression in human fibroblasts." *Crit Rev Eukaryot Gene Expr* **8**(1):43-80.
9. Cristofalo, V. J., Allen, R. G., Pignolo, R. J., Martin, B. G., and Beck, J. C. (1998b). "Relationship between donor age and the replicative lifespan of human cells in culture: a reevaluation." *Proc Natl Acad Sci USA* **95**(18):10614-10619.
10. Fleming, J. E., Reveillaud, I., and Niedzwiecki, A. (1992). "Role of oxidative stress in Drosophila aging." *Mutat Res* **275**(3-6):267-279.

11. Gensler, H. L., and Bernstein, H. (1981). "DNA damage as the primary cause of aging." *Q Rev Biol* **56**(3):279-303.
12. Larsen, P. L. (1993). "Aging and resistance to oxidative damage in Caenorhabditis elegans." *Proc Natl Acad Sci USA* **90**(19):8905-8909.
13. Longo, V. D., and Fabrizio, P. (2002). "Regulation of longevity and stress resistance: a molecular strategy conserved from yeast to humans?" *Cell Mol Life Sci* **59**(6):903-908.
14. Lu, T., Pan, Y., Kao, S. Y., Li, C., Kohane, I., Chan, J., and Yankner, B. A. (2004). "Gene regulation and DNA damage in the ageing human brain." *Nature* **429**(6994):883-891.
15. Martin, G. M., Austad, S. N., and Johnson, T. E. (1996). "Genetic analysis of ageing: role of oxidative damage and environmental stresses." *Nat Genet* **13**(1):25-34.
16. Mori, A., Utsumi, K., Liu, J., and Hosokawa, M. (1998). "Oxidative damage in the senescence-accelerated mouse." *Ann N Y Acad Sci* **854**:239-250.
17. Parrinello, S., Samper, E., Krtolica, A., Goldstein, J., Melov, S., and Campisi, J. (2003). "Oxygen sensitivity severely limits the replicative lifespan of murine fibroblasts." *Nat Cell Biol* **5**(8):741-747.
18. Sohal, R. S., Sohal, B. H., and Brunk, U. T. (1990). "Relationship between antioxidant defenses and longevity in different mammalian species." *Mech Ageing Dev* **53**(3):217-227.
19. Sohal, R. S., Mockett, R. J., and Orr, W. C. (2002). "Mechanisms of aging: an appraisal of the oxidative stress hypothesis." *Free Radic Biol Med* **33**(5):575-586.
20. Terman, A. (2001) "Garbage catastrophe theory of aging: imperfect removal of oxidative damage?" *Redox Rep.* **6**, 15-26.
21. Gallant, J. and Kurland, C. (1997) "The error catastrophe theory of aging. Point counterpoint." *Exp. Gerontol.* **32**, 333-337.
22. Gallant, J. and Parker, J. (1997) «The error catastrophe theory of aging. Point counterpoint.» *Exp. Gerontol.* **32**, 342-345.
23. Barja, G., and Herrero, A. (2000). "Oxidative damage to mitochondrial DNA is inversely related to maximum life span in the heart and brain of mammals." *Faseb J* 14(2):312-318.
24. Dice, J.F. and Goff, S.A. (1987) Error catastrophe and aging: future directions of research. In *Modern Biological Theories of Aging*. Warner, H.R., Butler, R.N., Sprott, R.L., and Schneider, E.L., Ed. Raven Press, New York. pp. 155-165.

25. Leonid A. Gavrilov and Natalia S. Gavrilova "The Reliability Theory of Aging and Longevity," *J. theor. Biol.* (2001) 213, 527-545.

26. Mattson, M. P., Duan, W., and Maswood, N. (2002). "How does the brain control lifespan?" *Ageing Res Rev* 1(2):155-165.

27. Boyko O. G. "Differentiation of radial glia cells into astrocytes is a possible ageing mechanism in mammals" // Zhurnal Obshchei Biologii (Journal of General Biology). V. 68 No 1 P. 35-51, 2007.

28. Harman, D. (1972). "The biologic clock: the mitochondria?" *J Am Geriatr Soc* 20(4):145-147.

29. Friguet, B., Bulteau, A. L., Chondrogianni, N., Conconi, M., and Petropoulos, I. (2000). "Protein degradation by the proteasome and its implications in aging." *Ann N Y Acad Sci* **908**:143-154.

30. Eriksson, M., Brown, W. T., Gordon, L. B., Glynn, M. W., Singer, J., Scott, L., Erdos, M. R., Robbins, C. M., Moses, T. Y., Berglund, P., *et al.* (2003). "Recurrent de novo point mutations in lamin A cause Hutchinson-Gilford progeria syndrome." *Nature* **423**(6937):293-298.

31. Marcel Leist, Anna F.Castoldi, *at all.(1997)* "Intracellular Adenosine Triphosphate (ATP) Concentration: A Switch in the Decision Between Apoptosis and Necrosis." Published April 21, 1997 // *JEM vol. 185 no. 8 1481-1486*

32. Michael Lutter, Guy A Perkins and XiaodongWang(2001) "The pro-apoptotic Bcl-2 family member tBid localizes to mitochondrial contact sites" Published: 8 November 2001 *BMC Cell Biology* 2001

33. Weismann, A. (1891). *On Heredity*. Claredon Press, Oxford.

34. Medawar, P.B. (1946) "Old age and natural death." *Modern Q.* **1**, 30-56.

35. Medawar, P. B. (1952). *An Unsolved Problem of Biology*. H. K. Lewis, London.

36. Miller, R. A. (1999). "Kleemeier award lecture: are there genes for aging?" *J Gerontol A Biol Sci Med Sci* **54**(7):B297-307.

37. Goldsmith, T. C. (2004). "Aging as an evolved characteristic-Weismann's theory reconsidered." *Med Hypotheses* **62**(2):304-308.

38. Gavrilov, L. A., and Gavrilova, N. S. (2002). "Evolutionary theories of aging and longevity." *ScientificWorldJournal* 2:339-356.

39. Keeley, J.E. and Bond, W.J. (1999) Mast flowering and semelparity in bamboos: the bamboo fire cycle hypothesis. *Am. Nat.* **154**, 383-391.

40. Friedman, D.B. and Johnson, T.E. (1988) "A mutation in the age-1 gene in *Caenorhabditis elegans* lengthens life and reduces hermaphrodite fertility." *Genetics*118, 75-86.
41. Friedman, D.B. and Johnson, T.E. (1988) "Three mutants that extend both mean and maximum life span of the nematode, *Caenorhabditis elegans*, define the age-1 gene." *J. Gerontol.* **43**, B102-B109.
42. Johnson, T.E. (1990) "Increased life-span of age-1 mutants in *Caenorhabditis elegans* and lower Gompertz rate of aging." *Science* 249, 908-912.
43. Lithgow, G.J., White, T.M., Melov, S., and Johnson, T.E. (1995) "Thermotolerance and extended life-span conferred by single-gene mutations and induced by thermal stress." *Proc. Natl. Acad. Sci. USA* **92**, 7540-7544.
44. Guarente, L. and Kenyon, C. (2000) "Genetic pathways that regulate ageing in model organisms." *Nature* **408**, 255-262.
45. B. J. Zwaan, R. Bijlsma and R. F. Hoekstra "On the developmental theory of ageing. I. Starvation resistance and longevity in *Drosophila melanogaster* in relation to pre-adult breeding conditions," *Heredity* (1991) 66, 29-39; doi:10.1038/hdy.1991.4.
46. Williams, G.C. (1957) Pleiotropy, natural selection and the evolution of senescence. *Evolution* **11**, 398-411.
47. Carlesworth, B. (1994) *Evolution in Age-Structured Populations.* Cambridge University Press, Cambridge.
48. Rose, M.R. (1991) *Evolutionary Biology of Aging.* Oxford University Press, New York.
49. Westendorp, R.G.J. and Kirkwood T.B.L. (1998) Human longevity at the cost of reproductive success. *Nature* **396**, 743-746.
50. Gavrilov, L.A. and Gavrilova, N.S. (1999) Is there a reproductive cost for human longevity? *J. Anti-Aging Med.* 2, 121-123.
51. Le Bourg, É. (2001) A mini-review of the evolutionary theories of aging. Is it the time to accept them? *Demogr. Res.* (Online) 4(1), 1-28.
52. Carnes, B.A. and Olshansky, S.J. (1993) Evolutionary perspectives on human senescence. *Popul. Dev. Rev.* 19, 793-806.
53. Campisi, J. (2001) From cells to organisms: can we learn about aging from cells in culture? *Exp. Gerontol.* 36, 607-618.
54. Itahana, K., Dimri, G., and Campisi, J. (2001) Regulation of cellular senescence by p53. *Eur. J. Biochem.* 268, 2784-2791.

55. Krtolica, A., Parrinello, S., Lockett, S., Desprez, P.-Y. and Campisi, J. (2001) Senescent fibroblasts promote epithelial cell growth and tumorigenesis: a link between cancer and aging. *Proc. Natl. Acad. Sci. U. S. A.*98, 12072-12077.

56. Economos, A.C. and Lints, F.A. (1986) Developmental temperature and life-span in *Drosophila-melanogaster*. 1. Constant developmental temperature—evidence for physiological adaptation in a wide temperature-range. *Gerontology*32, 18-27.

57. Partridge, L., Prowse, N., and Pignatelli, P. (1999) Another set of responses and correlated responses to selection on age at reproduction in *Drosophila melanogaster*. *Proc. R. Soc. London Ser. B* **266**, 255-261.

58. Hayflick, L., and Moorhead, P. S. (1961). "The serial cultivation of human diploid cell strains." *Exp Cell Res* **25**:585-621.

59. Hayflick, L. (1985). "The cell biology of aging." *Clin Geriatr Med* **1**(1):15-27.

60. Hayflick, L. (1994). *How and Why We Age*. Ballantine Books, New York.

61. Hayflick, L. (2000). "The future of ageing." *Nature* **408**(6809):267-269.

62. Hayflick, L. (2004). "Anti-aging" is an oxymoron." *J Gerontol A Biol Sci Med Sci* **59**(6):B573-578.

63. Artandi, S. E., Alson, S., Tietze, M. K., Sharpless, N. E., Ye, S., Greenberg, R. A., Castrillon, D. H., Horner, J. W., Weiler, S. R., Carrasco, R. D., *et al.* (2002). "Constitutive telomerase expression promotes mammary carcinomas in aging mice." *Proc Natl Acad Sci U S A* **99**(12):8191-8196.

64. Bassham, S., Beam, A., and Shampay, J. (1998). "Telomere variation in Xenopus laevis." *Mol Cell Biol* **18**(1):269-275.

65. Benard, C., and Hekimi, S. (2002). "Long-lived mutants, the rate of aging, telomeres and the germline in Caenorhabditis elegans." *Mech Ageing Dev* **123**(8):869-880.

66. Blackburn, E. H. (2000). "Telomere states and cell fates." *Nature* **408**(6808):53-56."

67. Blasco, M. A. (2005). "Telomeres and human disease: ageing, cancer and beyond." *Nat Rev Genet* **6**(8):611-622.

68. Collins, K., and Mitchell, J. R. (2002). "Telomerase in the human organism." *Oncogene* **21**(4):564-579.

69. Lindsey, J., McGill, N. I., Lindsey, L. A., Green, D. K., and Cooke, H. J. (1991). "In vivo loss of telomeric repeats with age in humans." *Mutat Res* **256**(1):45-48.

70. Dilman, V.M. "The Law of Deviation of Homeostasis and Diseases of Aging," John Wright PSG, Boston, 1981.

71. Dilman, V.M., and Dean, W. "The Neuroendocrine Theory of Aging and Degenerative Disease," The Center for Bio-Gerontology, Pensacola, Florida, 1992.

72. Bowen, R. L., and Atwood, C. S. (2004). "Living and dying for sex. A theory of aging based on the modulation of cell cycle signaling by reproductive hormones." *Gerontology* 50(5):265-290.

73. Kirkwood, T.B.L. "Evolution of Aging." Nature 270: 301-304, 1977.

74. Kirkwood, T.B.L. and Holliday, R. (1979) The evolution of ageing and longevity. *Proc. R. Soc. London Ser. B Biol. Sci.* 205, 531-546.

75. **Kirkwood, T.B.L. and Austad, S.N. (2000) Why do we age? *Nature* 408, 233-238**

76. **Darwin(32)Darwin, C. (1859) *On the Origin of Species by Means of Natural Selection, or, The Preservation of Favoured Races in the Struggle for Life*. J. Murray, London.**

77. Chris Venditti, Andrew Meade, & Mark Pagel "Phylogenies reveal new interpretation of speciation and the Red Queen," *Nature* **463**, 349-352 (21 January 2010)

78. Finch, C. E. (1990). *Longevity, Senescence, and the Genome*. The University of Chicago Press, Chicago and London.

79. **Domènec Farré, Nicolás Bellora, Loris Mularoni, Xavier Messeguer and M Mar Albà "Housekeeping genes tend to show reduced upstream sequence conservation** *"Genome Biology"* 2007, **8**:R140doi:10.1186/gb-2007-8-7-r140

80. Wray GA, Hahn MW, Abouheif E, Balhoff JP, Pizer M, Rockman MV, Romano LA **"The evolution of transcriptional regulation in eukaryotes."** *Mol Biol Evol* 2003, **20**:1377-1419. .

81. Tagle DA, Koop BF, Goodman M, Slightom JL, Hess DL, Jones RT **"Embryonic epsilon and gamma globin genes of a prosimian primate (***Galago crassicaudatus***). Nucleotide and amino acid sequences, developmental regulation and phylogenetic footprints."** *J Mol Biol* 1988, **203**:439-455.

82. Lenhard B, Sandelin A, Mendoza L, Engstrom P, Jareborg N, Wasserman WW **"Identification of conserved regulatory elements by comparative genome analysis."** *J Biol* 2003, **2**:13..

83. Dermitzakis ET, Clark AG **"Evolution of transcription factor binding sites in mammalian gene regulatory regions: conservation and turnover."** *Mol Biol Evol* 2002, **19**:1114-1121..

84. Alexandr Boldachev. "Novatsiy. Suzdenia v rusle evolutsionnoy paradigmy. // SPB.: Izd-vo S.—Peterb. un—ta, 2007.

85. Hutchison, C.A. III, and Montague, M.G.(2002). **Mycoplasmas and the minimal genome concept. In Molecular biology and Pathogenicity of Mycoplasmas.** pp. 221-253. Razin, S., and Herrmann, R. eds., Kluwer. Academic/Plenum Publishers

86. Maturana, Humberto R. and Varela, Francisco J. (1980) eds. Autopoiesis and Cognition-The Realization of the Living. Boston Studies in the Philosophy of Science, Volume 42. D. Reidel Publishing Company, Dordrecht, Holland. 141 pp.

87. Ni-Stor F. Gonzalez-Cadavid and Carmen Saez de Cordova **"Role ofMembrane-Bound and Free Polyribosomes in the Synthesis of Cytochrome c in Rat Liver,"** Biochem. J. (1974) 140, 157-167.

88. Bartke, A., Coschigano, K., Kopchick, J., Chandrashekar, V., Mattison, J., Kinney, B., and Hauck, S. (2001a). "Genes that prolong life: relationships of growth hormone and growth to aging and life span." *J Gerontol A Biol Sci Med Sci* **56**(8):B340-349.

89. Bartke, A. (2003). "Can growth hormone (GH) accelerate aging? Evidence from GH-transgenic mice." *Neuroendocrinology* **78**(4):210-216.

90. Masoro, E. J. (2005). "Overview of caloric restriction and ageing." *Mech Ageing Dev* **126**(9):913-922.

91. Piraino, Stefano; F. Boero, B. Aeschbach, V. Schmid (1996). "Reversing the life cycle: medusae transforming into polyps and cell transdifferentiation in Turritopsis nutricula (Cnidaria, Hydrozoa)." Biological Bulletin (Biological Bulletin, Vol. 190, No. 3) 190 (3):302-312. doi:10.2307/1543022

92. Michael Chorost "World Wide Mind The Coming Integration of Humanity, Machines, and the Internet" 2011, eBook 256p.

93. Ray Kurzweil "The Singularity Is Near: When Humans Transcend Biology" (2005)

www.ingramcontent.com/pod-product-compliance
Lightning Source LLC
Chambersburg PA
CBHW031327290526
45784CB00014B/2401